Is It Real?

Robert Anthony Levy

Prologue

At some point, people began asking a new kind of question.

Not whether something was true.

Not whether it existed.

But whether it was real.

The question appeared quietly. It slipped into everyday conversation without announcement or debate. It showed up on menus, in stores, in advertisements, and eventually in front of screens.

Is the food real.

Is the watch real.

Is this video real.

The question itself did not sound strange. That is what made it strange.

For most of human history, reality did not require confirmation. A thing was assumed to be what it appeared to be unless there was a reason to doubt it. Deception existed, but skepticism was situational, not automatic. People did not walk through the world auditing the authenticity of ordinary objects.

Something changed.

The change was not sudden and it was not caused by a single invention. It arrived gradually, as a habit rather than a decision. Over time, people learned that appearances could be separated from origins. That outcomes could be reproduced without reproducing the processes that created them. That the surface of a thing could remain convincing even when its substance had changed.

As this separation became common, trust quietly shifted.

Reality stopped being assumed and started being evaluated.

This did not mean the world became fake. It meant the burden of proof moved. Reality was no longer something you simply encountered. It was something you verified.

The question is it real is not philosophical. It is practical. It is asked not out of doubt, but out of competence. It reflects an understanding that modern life is filled with substitutions, reconstructions, optimizations, and representations that look correct while operating under different rules.

People ask the question because they have learned to.

Once learned, the habit does not stay confined to objects.

If food can be reconstructed, images can be staged, and materials can be simulated, then certainty itself begins to feel conditional. The question spreads from products to media, from media to stories, from stories to authority, and eventually inward.

Is this experience real.

Is this memory real.

Is this version of myself real.

The book that follows is about how this habit formed.

It is about the moment when reality stopped speaking for itself and began to arrive with explanations attached. About how trust moved away from substance and toward origin. About how meaning drifted from what something is to how it came to be.

And it is about where this habit leads.

Because once people learn to ask whether something is real, it becomes difficult to know where to stop asking.

By the end, one final question remains.

Not about an object.

Not about an image.

But about the thing you are holding right now.

Is this real.

Chapter 1 - The Question That Came First

The candles should have gone out.

They were small, set along a stone wall behind our table at an outdoor restaurant in Tulum. The night was warm and open. There was no glass, no enclosure, nothing separating us from the air. When the wind picked up, it did so without warning. It arrived in uneven bursts, strong enough to lift napkins and send ripples across the table.

The leaves moved.

Fabric shifted.

The heat from the flames wavered against my skin.

But the flames themselves did not.

They stayed upright and steady, unchanged by the air moving around them. No flicker. No hesitation. They burned as if the wind did not exist.

I watched them longer than I should have.

At first, I tried to make sense of it in ordinary ways. Perhaps the wall redirected the air. Perhaps the wind was stronger at table height than where the candles sat. Perhaps there was something I was missing.

That is usually where understanding begins.

But it is not where my mind went next.

Without effort, without intention, a different question appeared.

Are the candles real.

It was not a joke. It did not arrive as wonder or disbelief. It arrived cleanly and efficiently, as if it were the most reasonable thing to ask. The question required no buildup. It did not wait for confusion to resolve or curiosity to exhaust itself. It simply replaced the attempt to understand.

Only later did that bother me.

The candles were real. I still do not know exactly why they did not go out, but they did not. There was no reveal. No hidden mechanism. No explanation that restored the moment to comfort.

What stayed with me was not the behavior of the flames, but the behavior of my mind.

For most of human history, the natural response to that moment would have been different. A person might have leaned closer. They might have waited. They might have asked how. The assumption would have been that reality was intact, even if it was not yet understood.

My assumption was conditional.

Before I tried to understand what was happening, I questioned whether the thing itself belonged to reality at all.

The speed of that move was what unsettled me.

The question is it real did not arrive as skepticism. It arrived as fluency. As though my mind had learned that this was the correct first step. As though authenticity was something to verify before comprehension, not after.

The question did not feel philosophical. It felt practical. It felt trained.

At some point, many of us learned to think this way.

We live in a world filled with convincing substitutions. Light without fire. Flavor without ingredients. Texture without material. Images without events. Experiences without presence. These substitutions are not hidden. They are offered openly, often proudly, as improvements.

Over time, the mind adapts.

It learns that appearance no longer guarantees origin. That outcomes can be separated from the processes that once defined them. That what something looks like and what it is made of are no longer the same question.

This learning does not announce itself. It accumulates.

Eventually, the mind stops asking how first and starts asking what instead. Not what is happening, but what category does this belong to. Original or reproduced. Natural or engineered. Captured or generated.

Is it real.

What makes this habit difficult to notice is that it feels like intelligence. It feels like sophistication. It feels like keeping up.

And often, it is useful.

But usefulness has side effects.

Once the question is installed, it does not wait politely for evidence of deception. It appears preemptively. It steps in before curiosity has finished its work. It becomes the default lens through which experience is filtered.

That night in Tulum, nothing false was revealed. Reality did not fail. My perception did not collapse.

But something subtle shifted.

I realized that the question had arrived too early. That it had interrupted understanding rather than followed it. That my first instinct was not to engage with the world, but to audit it.

That realization lingered long after the candles stopped mattering.

If the mind now questions reality before it questions explanation, then the change is not about technology alone. It is about posture. About where trust is placed. About what feels safe to assume.

This book is not an argument that the world has become fake. It is an attempt to understand why authenticity has become a prerequisite rather than a conclusion.

It asks when the question is it real became automatic, what conditions taught us to ask it, and what quietly changes when we do.

Because once a mind learns to ask that question first, it becomes difficult to remember how it ever did anything else.

And once that habit spreads from objects to images, from images to stories, and from stories to experience itself, the question stops being about reality alone.

It becomes a way of relating to the world.

Chapter 2 - Verification Becomes Cheap

Verification Becomes Cheap

There was a time when most things did not need to be verified. Not because people were naïve, but because the cost of verification was higher than the cost of trust. You believed what you saw because seeing was usually enough. You believed the person in front of you because they were present. You believed objects were what they appeared to be because there were few incentives and few mechanisms to convincingly fake them.

This was not a world without deception. It was a world where deception required effort, proximity, and risk. To fake something convincingly, you had to be there. To lie persuasively, you had to sustain it over time. The friction of the physical world acted as a filter. It limited scale. It limited repetition. It limited reach.

Verification existed, but it was heavy. It required specialists, time, or authority. Most people rarely encountered situations where verification felt necessary, because most interactions were local, slow, and human.

That balance changed when verification became cheap.

Cheap does not mean inaccurate. It means easy. It means fast. It means available on demand. When verification becomes cheap, it stops being an event and becomes a habit. And when it becomes a habit, it reshapes how trust works.

At first, this feels like progress. It is framed as empowerment. You can check. You can confirm. You can look it up. You no longer have to rely on anyone else. The world becomes searchable, comparable, and sortable.

But something subtle happens at the same time. The act of verification starts to replace the act of perception.

Instead of asking what something is, we ask whether it has been validated elsewhere. Instead of trusting our senses, we trust systems that claim to measure authenticity. Instead of forming judgment through experience, we defer judgment to markers that promise certainty.

The question "is it real?" begins to surface not because reality has become rarer, but because confirmation has become effortless.

Consider how this shows up in ordinary life.

A diamond no longer announces itself by appearance alone. To most people, a lab-grown diamond looks identical to a mined one. The difference lives in paperwork, certification, and provenance. Without a grading report or a trusted seller, the object itself does not resolve the question. Verification is required, and it is readily available. So the question is asked automatically.

Beef labeled as wagyu does not demand belief based on taste alone. It arrives with branding, pricing, and claims of origin. The consumer is trained to look for labels, not flavor. The question is not whether it tastes good, but whether it qualifies.

A hardwood floor is no longer assumed to be wood. It might be engineered. It might be a veneer. It might be synthetic. The surface deceives the eye by design. Authenticity lives beneath the surface, invisible without disclosure.

A designer handbag may be visually indistinguishable from a counterfeit. The difference resides in serial numbers, stitching patterns, receipts, or a sales channel deemed trustworthy. Without those signals, the object is suspended between categories.

Seafood provides another example. Crab might be imitation. Tuna might be treated. Shrimp might be farmed under conditions that change what the label implies. The consumer rarely verifies directly. Instead, they rely on menus, certifications, or brand reputation. The question "is it real?" becomes shorthand for "has someone else already checked this for me?"

Gold raises the same issue. Its weight, color, and feel once sufficed for everyday certainty. Today, purity is assumed only with testing, stamping, or institutional backing. The object itself no longer settles the matter.

In each case, verification is available with minimal effort. A search. A scan. A certificate. A logo. A trusted intermediary.

What matters is not whether these systems are good or bad. What matters is that they have trained us to experience reality indirectly.

When verification becomes cheap, trust is no longer built through exposure. It is outsourced to mechanisms designed to scale.

This changes behavior.

People stop learning how to tell. They learn how to check.

The skill shifts from perception to navigation. From judgment to lookup. From experience to reference.

This does not make people less intelligent. It makes them differently trained.

Over time, the body of knowledge that once lived in senses and memory migrates outward. It becomes encoded in systems, standards, and institutions. These systems are efficient. They reduce error. They allow society to function at scale.

But they also create distance.

The more we rely on verification, the less comfortable we become with ambiguity. The less tolerant we become of uncertainty. The less willing we are to sit with incomplete information.

So the question "is it real?" starts to appear even when the object itself has not changed.

This is why people sometimes ask the question even when nothing is fake.

A candle that does not blow out in the wind may still be real. The question arises not because the candle is artificial, but because the experience violates expectation. The mind, trained to verify, reaches for an explanation that lives outside direct perception.

The important shift is not that people doubt more. It is that doubt has become procedural.

It is a reflex.

Instead of trusting and then correcting when wrong, we verify first and experience second.

This inversion has consequences.

Reality begins to feel provisional. Objects feel incomplete until validated. Experiences feel unfinished until confirmed by an external signal. Even personal reactions are sometimes deferred until others have responded.

The world becomes a series of claims awaiting verification.

And because verification is cheap, it is used constantly. Not always because it is needed, but because it is available.

This is the moment where "is it real?" stops being a practical question and starts becoming a cultural posture.

It is no longer limited to cases of risk or fraud. It migrates into everyday life. Into food. Into objects. Into images. Into stories. Into identity.

Once verification is expected, trust becomes conditional.

Not because people are suspicious, but because systems have taught them that certainty is always one step away.

The irony is that the more verification we introduce, the less grounded we feel. Certainty becomes dependent rather than embodied. Confidence is borrowed rather than built.

This chapter is not an argument against verification. It is an observation about cost.

When verification was expensive, trust was the default. When verification became cheap, trust became optional.

The question "is it real?" did not emerge because reality failed.

It emerged because confirmation became effortless.

And once a society learns to ask that question automatically, it never fully unlearns it.

Chapter 3 - Frictionless Doubt

There was a time when doubt carried weight.

To question something meant risking embarrassment, conflict, or consequence. To ask whether a claim was true was to imply that it might not be. To express uncertainty was to reveal something about yourself: your knowledge, your courage, your willingness to stand apart.

Doubt was not casual. It was a posture.

Because it had cost, it tended to resolve. You either accepted a claim or rejected it. You either believed or you did not. Uncertainty could not be held indefinitely without becoming uncomfortable.

That discomfort forced decisions.

Something changed when verification became easy.

Not easy in the sense of simple, but easy in the sense of private, instant, and consequence free. A question could be asked without being spoken. A doubt could be entertained without being expressed. A claim could be checked without anyone knowing it had ever unsettled you.

This altered the shape of doubt itself.

Today, questioning does not require commitment. It does not require belief or disbelief. It barely requires attention.

You can wonder without caring.

You can check without concluding.

You can ask without standing anywhere.

A person can look something up not to learn it, but to quiet a feeling. Not because the answer matters, but because uncertainty briefly surfaced and needed to be smoothed over.

The act is closer to reassurance than inquiry.

This is how doubt becomes frictionless.

It slips into the mind without effort and leaves without residue. It does not demand resolution. It does not insist on coherence. It simply passes through, leaving the surface undisturbed.

You sense something is off.

You glance for confirmation.

You move on.

The question is asked, but it is not held.

This would have been difficult in earlier worlds. When checking took time, effort, or social exposure, doubt lingered. It had to be reckoned with. It either grew into skepticism or collapsed into trust.

Now it evaporates.

Because doubt no longer carries cost, it no longer carries weight. It does not sharpen belief. It does not harden disbelief. It exists in a suspended state, never fully forming into either.

This changes how belief itself behaves.

Beliefs used to feel like positions. They were places you stood. They shaped how you spoke, how you acted, how you aligned yourself with others.

Now belief often feels provisional. Temporary. Revocable at any moment.

You believe something until you don't.

You accept something until it feels slightly wrong.

You move on without needing to reconcile the change.

This is not cynicism. It is flexibility taken to an extreme.

The mind learns that it does not need to commit because commitment is no longer required to function. You can operate effectively while holding everything lightly. You can navigate the world without resolving most questions that pass through your awareness.

In many ways, this is efficient.

But efficiency has side effects.

When nothing demands commitment, meaning thins. When belief never solidifies, it never deepens. When doubt never costs anything, it never clarifies.

Everything remains adjustable.

Nothing fully settles.

This creates a peculiar emotional texture.

People are not deeply suspicious, but they are rarely convinced. They are not hostile to claims, but they are not anchored to them either. They live in a state of soft uncertainty, constantly updated, constantly revised.

Reality becomes something you skim rather than inhabit.

The question is it real appears in this space not as an accusation, but as a reflex. It arises automatically, without urgency, without expectation.

Often the answer does not matter.

The asking is enough.

It momentarily restores a sense of orientation. It reassures the mind that it is still capable of questioning, even if it does not follow through. It is a gesture toward agency rather than an act of it.

This is the paradox of frictionless doubt.

The easier it becomes to question reality, the less impact questioning has.

The question survives, but its force dissolves.

We have become fluent in uncertainty without being transformed by it. We touch doubt lightly and move on. We question without standing. We verify without believing.

This is not a failure of intelligence. It is an adaptation.

The mind has learned how to live comfortably in a world where certainty is always provisional and rarely required. It has learned to regulate itself through micro checks, quick confirmations, and fleeting reassurances.

But something is lost in the process.

When doubt no longer sharpens belief, belief no longer organizes experience. When nothing settles, nothing fully means.

This is the condition we carry into the next question.

Not whether something is real.

But what reality feels like when it never quite lands.

Chapter 4 - The Age of "Real"

At some point, the word real stopped being descriptive and became necessary.

It began appearing in places where it had never been required before. On menus. On labels. In showrooms. In ordinary conversation. People started asking questions that would once have sounded unnecessary, even odd, despite the fact that the objects in question were plainly present.

Is the shrimp real.

Is the tuna real.

Is this real wood.

Is it real leather.

These questions are not abstract or ironic. They are practical. They are asked calmly, often without suspicion. And yet they signal a quiet change in how the world is understood.

When someone asks whether the shrimp is real, they are not questioning its existence. The shrimp is visible and tangible. The question is whether it is what it claims to be. Whether it has been substituted, reconstructed, diluted, or optimized in a way that breaks the promise carried by its name.

The same is true of wood, leather, stone, fabric, and metal. These materials still surround us. But their presence no longer guarantees their identity. Appearance no longer certifies origin.

Reality has become conditional.

This shift did not occur because people suddenly grew distrustful. It occurred because systems became powerful.

For most of human history, objects carried their origins openly. Wood revealed grain. Leather announced itself by smell and feel. Food spoiled quickly and visibly. Craftsmanship left marks that could not be replicated cheaply or at scale. The distance between source and use was short enough that authenticity rarely required explanation.

Industrialization stretched that distance. Mass production blurred it. Global supply chains erased it almost entirely. Objects began arriving without context, detached from place, optimized for consistency and cost rather than story.

Once that happened, origin could no longer be assumed. It had to be stated.

This is why labels multiplied. Real butter. Genuine leather. Solid wood. Authentic stone. These phrases did not emerge from philosophical curiosity. They emerged because imitation had become capable enough to require distinction.

The word real became a signal.

Not of existence, but of trust.

Food makes this especially clear.

Certain foods are routinely substituted or relabeled. This does not require dramatic deception. It requires incentives. When food becomes abstracted into logistics, margins, and efficiency, the eater adapts. The question shifts from whether something will nourish you to whether it aligns with the expectation attached to its name.

Is this what it says it is.

That question does not stay contained.

Furniture is no longer assumed to be made of what it resembles. Floors that look like wood are composites. Tables that feel heavy may be hollow. Stone is printed. Metal is layered. The surface no longer testifies to the substance beneath it.

The question follows quietly and naturally.

Is it real.

Jewelry sharpens the problem further.

For most of history, asking whether a gem was real meant asking whether it was stone or glass. The distinction mattered because the materials behaved differently over time. Reality could be inferred through use, wear, and durability.

Today, the question has changed.

A lab grown diamond is chemically identical to a mined one. It refracts light the same way. It lasts just as long. Specialized equipment is required to tell them apart. In every material sense, it is real.

And yet the question remains.

What is being asked is no longer about composition. It is about origin, scarcity, and story. It is about whether the object participates in the same social meaning as its counterpart.

Here, the word real begins to fail.

When something can be materially indistinguishable and still feel uncertain, reality alone is no longer sufficient. Meaning has shifted away from substance and toward narrative.

This is not a collapse of honesty. It is a consequence of capability.

Modern systems can reproduce outcomes without reproducing history. They can deliver appearance without context. The result performs as expected but occupies an ambiguous place in the world. The object is real, but its meaning is unsettled.

So the consumer adapts.

The question becomes less about existence and more about belonging.

Does this align with what I think it represents.

Does it fit the category I assume it belongs to.

Does it follow the rules I expect.

The phrase is it real becomes shorthand for this entire evaluation.

What is striking is how quickly this habit spreads.

Once people learn to question materials, they begin to question images. Once images are questioned, stories follow. Once stories are questioned, authority weakens. Once authority weakens, certainty erodes.

Reality becomes something that arrives with conditions.

This is not paranoia. It is learned behavior.

In a world where imitation is often indistinguishable from origin, where optimization replaces intention, and where appearance has been separated from process, asking whether something is real is not cynicism. It is competence.

But competence has a cost.

When reality must announce itself, the announcement becomes part of the construction. When authenticity must be declared, it becomes performative. When trust must be signaled, it becomes fragile.

Quiet reality no longer feels sufficient. Loud reality feels safer, even when the volume itself is manufactured.

This is the age of real.

An age in which reality still exists, but no longer speaks for itself.

An age in which existence is insufficient and meaning depends on explanation.

And once reality requires explanation, it becomes possible for explanation to replace it entirely.

Chapter 5 - Seeing and Believing

For most of human history, seeing was not believing. It was only seeing.

Vision was intimate and limited. You trusted what you saw because it happened in front of you, in your presence, within the same space your body occupied. Distance weakened certainty. Reports traveled through voices, not images. Stories changed as they moved. Authority mattered more than depiction.

An image was not evidence. It was interpretation.

This is difficult to remember now.

Photography changed that relationship almost instantly. Not because photographs were perfectly truthful, but because they felt unarguable. Light itself appeared to be doing the work. No narrator stood between the event and the viewer. No storyteller shaped the account. The world seemed to imprint itself directly onto the surface.

For the first time, something could be seen without being witnessed.

This was not merely a new technology. It was a new expectation.

Images began to arrive detached from experience. A moment could be transported intact, stripped of context, intention, and consequence. The viewer was not present when it happened, yet felt entitled to certainty.

Seeing started to carry weight.

Early photographs did not initially provoke suspicion. They provoked awe. People gathered around them as proof that the world could now be captured. That reality could be frozen and carried elsewhere. That truth could be transported.

But something subtle followed.

Once images could stand in for presence, the question of authenticity could not be far behind.

Was the image altered. Was the scene staged. Was the moment chosen to mislead. Was this what it looked like before or after something happened.

None of these questions would have made sense earlier. They only arise once representation replaces experience.

Film intensified this shift. Motion brought credibility. Time unfolding on screen felt undeniable. Yet even as audiences gasped at early cinema, they were learning a new habit without realizing it.

They were learning to watch while withholding full belief.

Editing existed. Framing mattered. Scenes could be constructed. Meaning could be guided. The image no longer simply showed. It suggested.

Seeing became something you did carefully.

This is where the modern posture toward reality began to form. Not suspicion, exactly, but distance. A quiet gap between perception and acceptance.

Images no longer ended questions. They initiated them.

Is this staged. Is this selective. Is this the whole story.

The image demanded verification.

Once that expectation existed, it never went away.

Today, images arrive constantly, detached not only from experience but from effort. They require no travel, no presence, no patience. They appear instantly, perfectly framed, endlessly replayable.

And yet they rarely settle the mind.

A photo without context feels incomplete. A video without provenance feels unstable. A clip without confirmation feels provisional.

Seeing is no longer enough.

This is not because images became worse. It is because we learned how easily they could stand in for reality. And once substitution is possible, certainty becomes conditional.

The phrase is it real begins to make sense here.

It does not accuse the image of lying. It acknowledges that seeing has been separated from knowing.

The question appears not because the image failed, but because the relationship changed.

What once felt like proof now feels like a prompt.

This habit extends far beyond photographs and film. It quietly shapes how we encounter everything presented visually. Screens, feeds, interfaces, diagrams, dashboards. The world now arrives prepackaged for viewing, and viewing no longer implies trust.

Reality has learned how to present itself convincingly. The mind has learned how to hesitate politely.

This is not a loss of intelligence. It is a learned adaptation.

We are living downstream from the moment when images stopped requiring witnesses and started requiring confirmation.

Once that happened, belief could no longer rest on sight alone. It needed reinforcement. Context. Agreement. Consensus.

Seeing became the beginning of the question, not the end of it.

And so, even now, when something appears entirely ordinary, entirely plausible, entirely well formed, the reflex surfaces quietly.

Is this real.

Not because the eyes failed.

But because seeing no longer guarantees anything by itself.

Chapter 6 - Making and Imitation

Once seeing became separable from believing, owning could not remain untouched.

For most of human history, objects were difficult to imitate. Materials were local. Skills were unevenly distributed. Craft left fingerprints. Even when copies existed, they announced themselves through small imperfections. The real thing did not need defending. It was simply better.

Authenticity was not a category. It was assumed.

Industrial production disrupted that assumption quietly at first. Machines did not get tired. They did not drift. They did not leave telltale signs of human variance. What they produced was consistent in a way handwork never was.

Consistency began to look like quality.

This reversal matters.

When copies improve, authenticity stops being self evident. When imitation becomes convincing, reality needs explanation.

At first, mass production was celebrated. It promised access, affordability, and sameness. The same chair. Thc same plate. The same garment. The same experience. Equality through repetition.

But repetition has a side effect.

When many things look identical, the question shifts from how well it works to what it is.

Is this the original.

Is this licensed.

Is this the real one.

These questions would have sounded strange in earlier eras. A tool was real if it worked. A garment was real if it kept you warm. A table was real if it held weight.

Meaning lived in function.

As imitation improved, meaning migrated.

The object itself became less important than its story. Where it came from. Who made it. What it signified. Whether it belonged to an accepted lineage.

This is when authenticity became a property that had to be asserted.

Labels appeared. Serial numbers. Certificates. Brands that promised continuity across copies. Marks that reassured the buyer that this instance was connected to something larger than itself.

The object no longer spoke for itself. It needed a narrator.

This did not happen because people became gullible. It happened because the surface stopped being reliable.

Two things could look identical and not mean the same thing. One could carry history, labor, and status. The other could carry only appearance.

Once that distinction exists, it never disappears.

The modern world is saturated with objects that perform correctly while remaining questionable. They function. They endure. They satisfy practical needs. And yet they provoke a quiet hesitation.

Is this solid wood or veneer.

Is this handmade or manufactured.

Is this gold or plated.

The questions arise not because the objects fail, but because success no longer guarantees authenticity.

Ownership becomes interpretive.

To own something now is not just to possess it, but to understand what it represents. To know whether it belongs in the category you think it does. To be able to explain it if asked.

Even when no one asks.

This creates a new burden.

The owner becomes responsible not only for the object, but for its verification. The meaning of the thing rests partly on the owner's ability to place it correctly within a system of distinctions.

Reality, once again, is delegated.

You trust that someone else has already sorted originals from copies. That standards exist. That enforcement happens somewhere out of sight. That the object you hold has passed through the right channels.

And when those assurances are present, the question quiets.

When they are absent, it returns.

Is it real.

Not because the object feels fake.

But because function no longer settles the matter.

This shift reshapes desire itself.

People begin to want not just things, but confirmation. Not just quality, but legitimacy. Not just possession, but assurance that possession means what they think it means.

The more convincing imitation becomes, the more energy is spent distinguishing it from the real.

This is not a failure of taste. It is a structural consequence of abundance.

When making becomes easy, meaning becomes fragile.

And so the question follows the object everywhere.

It lingers in showrooms and closets, in hands and in minds. It appears even when the thing performs perfectly, even when it satisfies every practical requirement.

Because performance is no longer the point.

The world learned how to make things that look real.

The mind learned how to ask whether looking real is enough.

Chapter 7 - Eating Reality

Food used to be the most difficult thing to fake.

It spoiled quickly. It came from nearby. It carried the marks of season, soil, and hand. You knew what you were eating because you watched it grow, raise, or arrive. Taste was memory. Texture was evidence. The body learned what was real by repetition.

To eat was to trust something intimately.

That intimacy did not survive scale.

As food systems expanded, distance entered the meal. Ingredients traveled farther. Processes lengthened. Preservation improved. The connection between source and plate stretched until it thinned.

Food still arrived reliably. It became safer, more abundant, more consistent. But it also became less legible.

You could no longer tell what something was by tasting it alone.

This is where narration entered the meal.

Menus began to explain. Packaging began to reassure. Words appeared where experience once sufficed. Organic. Wild. Farm raised. Grass fed. Authentic. Traditional.

Food started telling stories about itself.

These stories did not arise because people suddenly became suspicious. They arose because the sensory loop had been broken. The body no longer had enough information on its own.

Two dishes could taste identical and mean entirely different things. Two cuts of meat could look the same and carry opposite values. One might signify care, origin, and restraint. The other efficiency, scale, and anonymity.

Taste could not resolve the difference.

So the question moved upstream.

Where did this come from.

What is it really made of.

Is this the real thing.

Eating became interpretive.

You no longer consumed food alone. You consumed its explanation. The meal included context. The act of eating quietly asked whether what you were ingesting matched the story attached to it.

This is why authenticity in food feels unusually charged.

You are not just buying calories. You are incorporating something into your body. The distance between the claim and the consequence is small. When reality feels uncertain here, discomfort rises quickly.

Is this fish actually what the menu says.

Is this oil what it claims to be.

Is this flavor natural or constructed.

These questions would have seemed excessive in earlier eras. Hunger settled most debates. Nourishment was enough.

Now nourishment is assumed. Meaning is not.

Food becomes a site where delegated reality is felt viscerally. You trust that inspections happened. That standards exist. That the story was vetted somewhere you cannot see.

And yet, the question persists.

Because food enters the body, not just the mind.

The modern eater lives inside a contradiction. They are surrounded by abundance and haunted by uncertainty. They have more choice than ever and less sensory authority than before.

So they read. They ask. They scan. They learn which words reassure them and which unsettle them. They develop shortcuts for trust.

They also develop fatigue.

Eventually, the question is it real stops demanding answers. It becomes a background hum. Something you register and then set aside in order to eat at all.

This is not failure. It is adaptation.

The body still needs food. The mind still seeks meaning. The system cannot provide certainty at scale. So the individual learns to live with partial assurance.

Eating becomes an act of managed trust.

You accept the story well enough to proceed. You suspend deeper doubt because the meal must continue.

And yet, the question never fully disappears.

It waits quietly behind the menu, behind the packaging, behind the language of authenticity that now accompanies even the most ordinary meal.

You taste.

You swallow.

You wonder.

Is this real.

Not because it isn't nourishing.

But because nourishment no longer tells the whole story.

Chapter 8 - When Reality Became a Brand

At some point, reality stopped being something we encountered and became something we purchased.

This did not happen all at once. There was no announcement. No cultural vote. It happened quietly, through packaging, labels, certifications, and marketing language that promised reassurance more than truth.

"Authentic."

"Genuine."

"Real."

These words did not exist to describe reality. They existed to calm doubt.

A tomato did not need to be called real until something resembling a tomato appeared beside it. Beef did not need lineage until alternatives emerged. Gold did not need testing until appearances became cheap.

Reality did not become fragile. Our confidence in it did.

Brands learned this faster than institutions. They understood that what people wanted was not proof, but relief. Not certainty, but comfort. A symbol that allowed the mind to move on without friction.

A label could now do the work that knowledge once did.

We began outsourcing judgment to packaging.

This is when "real" stopped being a property of an object and became a feature of persuasion.

Organic.

Artisanal.

Handcrafted.

Heritage.

Farm-raised.

Conflict-free.

Natural.

None of these words guarantee anything on their own. They operate emotionally, not empirically. They are designed to close the question, not answer it.

They say: you don't need to think anymore.

And that is their power.

Reality, once branded, became scalable.

A single farm could feed millions. A single factory could produce thousands of "handmade" items. A single image could stand in for experience. Authenticity could now be replicated, even when the thing it described could not.

This was not deception in the crude sense. It was optimization. It was the industrialization of reassurance.

We didn't want to know how something was made. We wanted to know that it was safe to stop asking.

And brands delivered.

The modern consumer does not distrust reality. They distrust their ability to evaluate it. So they look for shortcuts. Signals. Proxies.

A logo.

A certification stamp.

A price point.

These are not proofs. They are social agreements. Shared myths that say, "Enough people believe this for you to accept it."

Reality became democratic in appearance and centralized in control.

The irony is that the more options we were given, the less direct contact we had with anything itself. We interacted with layers. Claims about claims. Stories about origins. Narratives designed to feel older, simpler, truer than the system that produced them.

We began consuming the idea of reality instead of reality.

This is why counterfeit goods feel so unsettling. Not because they fail, but because they succeed too well. They reveal that what we trusted was not the object, but the story wrapped around it.

If a fake handbag performs identically, what exactly was real about the original?

If lab-grown diamonds refract light the same way, what does authenticity mean?

If plant-based meat satisfies the same appetite, what is being preserved when we insist on the "real" version?

These questions are uncomfortable because they expose something deeper.

We don't fear fake things.

We fear that reality itself has become negotiable.

Brands did not create this fear. They responded to it.

They learned that selling certainty was more profitable than selling truth. That "real" did not need to be defined, only invoked.

So reality became a promise instead of an experience.

And once reality becomes a promise, it becomes something that can be broken.

This is the moment when "Is it real?" stopped being philosophical and became transactional.

Not a question about existence.

A question about value.

We were no longer asking whether something existed in the world. We were asking whether it justified its price, its story, its place in our lives.

Reality had entered the marketplace.

And it was no longer free.

Chapter 9 - What We Own Becomes Who We Are

Ownership used to be practical.

You owned what you needed to survive, to work, to shelter yourself from the world. Objects were tools. They extended your capability, not your identity.

Somewhere along the way, ownership changed its job.

What we own began to speak for us.

Not loudly. Subtly. Through signals that required no explanation. A watch on the wrist. Shoes on the feet. A logo on a bag. A material in a kitchen. A car in a driveway.

These objects stopped answering the question "what does this do?" and started answering "who is this for?"

Identity moved into the marketplace.

This is where "is it real?" took on its most personal form. Not as a question about the object, but about the person holding it.

Is the watch real?

Is the bag real?

What is really being asked is something else.

Is your story real?

Because when ownership becomes symbolic, authenticity becomes moral. A fake object is no longer just a lesser object. It becomes a lie.

This is why counterfeit goods provoke such strong reactions. Not because they fail mechanically, but because they succeed socially.

They blur the boundary between earned identity and performed identity.

If someone cannot tell the difference, what exactly is being protected?

The discomfort is not about materials. It is about meaning.

Luxury learned this early. Scarcity was no longer about supply. It was about separation. Price became a filter. Not to guarantee quality, but to restrict access to identity.

To own the thing was to belong to a category.

Realness became a gate.

This logic spread quietly into everyday life.

Real hardwood floors.

Real leather seats.

Real gold jewelry.

These objects promise durability, but they also promise legitimacy. They say you chose correctly. You invested wisely. You did not settle.

Ownership became a résumé written in objects.

At the same time, replication improved. Alternatives became indistinguishable. Lab diamonds sparkled. Engineered wood endured. Synthetic fabrics outperformed natural ones.

The functional difference shrank. The symbolic difference did not.

So the question shifted again.

If the experience is the same, what exactly am I paying for?

The answer is rarely spoken out loud.

You are paying for narrative continuity. For the reassurance that your identity aligns with the story you want to tell about yourself. For the ability to stop wondering whether you are pretending.

This is why people ask for receipts. Certificates. Provenance. Documentation.

They are not protecting the object. They are protecting the self.

Ownership now carries psychological weight it was never designed to bear. Objects have become mirrors. And mirrors are unforgiving.

If something you own is revealed to be fake, it does not feel like an error. It feels like exposure.

This is the collision point. Where reality, value, and identity meet.

We no longer ask "is it real?" to understand the world. We ask it to understand where we stand within it.

And once identity enters the equation, doubt becomes dangerous.

Because if the object is not real, maybe the status was borrowed. Maybe the confidence was fragile. Maybe the line between having and being was thinner than we thought.

So we defend realness aggressively. We police it socially. We mock knockoffs. We draw lines.

Not to preserve truth, but to preserve meaning.

Ownership has become performance.

And authenticity has become a role we feel pressured to play convincingly.

This is the quiet tragedy of modern reality. Objects were never meant to carry this much existential weight. They were never meant to answer questions about worth.

But we handed them that responsibility anyway.

And now, every purchase whispers a question we can no longer avoid.

Is this real enough to be me?

Chapter 10 - When Reality Turned Inward

For most of human history, reality was external.

It was the weather, the land, the body, the visible actions of others. Reality resisted us. It pushed back. It did not ask for interpretation.

Then, slowly, reality moved inside.

This shift did not begin with philosophy. It began with tools. Screens. Interfaces. Systems designed to personalize experience. What you saw became different from what others saw. What you knew became curated. What felt real became increasingly internal.

Reality stopped being shared by default.

Once this happened, the question "is it real?" changed one final time.

It was no longer about objects.

No longer about images.

No longer about ownership.

It became about perception itself.

If two people experience the same event differently, which version is real? If memory edits itself over time, was the original moment ever stable? If emotion reshapes facts, does truth survive feeling?

We entered an era where reality was no longer something you encountered, but something you assembled.

Algorithms learned your preferences faster than you did. They filtered the world until it felt familiar. Comfortable. Confirming. Reality began to feel right even when it was incomplete.

And because it felt right, it felt real.

This is the quiet power of personalization. It does not lie overtly. It selects. It removes friction. It presents a version of the world that agrees with you often enough that you stop questioning it.

Doubt becomes inefficient.

This is why conspiracy feels persuasive. Why simulations feel plausible. Why people ask whether events, identities, even existence itself might be constructed.

It is not because people have lost their grip on reality. It is because reality no longer arrives whole.

We live inside partial truths that feel complete.

At the same time, technology blurred the boundary between thought and expression. Ideas could be externalized instantly. Images generated. Voices replicated. Text produced.

The internal world gained the ability to masquerade as the external one.

This is the moment where the final inversion occurs.

Instead of asking whether the world is real, we begin asking whether our experience of it is.

Is this memory real?

Is this feeling justified?

Is this belief mine, or was it planted?

Reality becomes psychological.

And once reality is psychological, it becomes fragile.

Not because it is false, but because it is personal.

Two people can now live in entirely different realities while sharing the same physical space. They are not pretending. They are not confused. They are responding to different inputs, different narratives, different reinforcements.

Reality fractures quietly.

The question "is it real?" no longer seeks an answer. It seeks reassurance. It asks whether the ground beneath us is stable enough to stand on.

This is why the question has spread everywhere. Food. News. Identity. History. Relationships. Even the self.

We are not becoming more skeptical. We are becoming more aware of how much interpretation reality now requires.

The strange truth is this.

Reality did not disappear.

It multiplied.

And navigating it now demands something previous generations did not need as urgently.

Humility.

The humility to admit that what feels real may be incomplete. That certainty is no longer a guarantee of truth. That confidence does not equal accuracy.

The future will not be decided by who controls reality, but by who can live inside uncertainty without retreating into illusion.

The most important version of "is it real?" is no longer a challenge directed outward.

It is a quiet question we ask ourselves.

Am I seeing clearly?

Am I choosing thoughtfully?

Am I mistaking comfort for truth?

Reality, in the end, is not something we verify once and move past.

It is something we negotiate continuously.

And that negotiation may be the most real thing left.

Epilogue - The Question Returns

This book began with a question that seemed small.

Is it real?

Along the way, that question widened. It attached itself to objects, images, brands, ownership, identity, and finally perception itself. It became less about verification and more about reassurance. Less about truth and more about comfort.

Somewhere in that widening, the question turned inward.

I should say something plainly now.

The idea for this book is mine. The questions are mine. The unease that led to them is mine. But the words you just read were not produced by my mind alone.

I am a non-linear thinker living in a mostly linear world.

My mind does not move step by step. It sees systems all at once. Conclusions arrive before explanations. I understand wholes in flashes, but struggle to sequence them in a way that makes sense to people who move comfortably from sentence to sentence, paragraph to paragraph.

For most of my life, this felt like a handicap.

I thrived in logic and mathematics, where structure could be inferred rather than narrated. But language required order. It demanded that ideas wait their turn. That they line up neatly, one after another, instead of arriving as a simultaneous flood.

Growing up in Canada, this gap became obvious early. To graduate high school, I needed a French credit. I failed repeatedly. Not because I lacked effort, but because my brain resisted linear expression in a language that required precision and sequence.

It wasn't until Grade 11 that a private tutor helped me obtain the credit I had been unable to earn on my own. I passed. Or perhaps she felt sorry for me and gave me the credit.

Either way, the tension remained.

Clear thinking trapped behind imperfect words.

This time, I chose a different approach.

I used artificial intelligence to help translate non-linear thought into linear language. Not to invent the ideas, but to sequence them. To take a system I could see instantly and unfold it in a way another mind could follow.

Does that make this book less real?

The experiences are mine. The observations are mine. The confusion that sparked the question is genuine. The collaboration is technological.

Throughout this book, we've seen how often humans outsource judgment. How easily we accept proxies for truth. How comfortable we are letting systems handle verification so we can move on.

And now, here you are, holding a book that openly admits it was written with the help of a machine.

If the ideas resonated, where did that resonance come from?

From the words?

From the mind behind them?

From your own recognition as you read?

Reality, as we've seen, no longer arrives whole. It is assembled. Negotiated. Shared between people, tools, and context.

This book exists. You read it. Something happened inside you while you did.

And that brings us to the final question.

Is this book real?

Made in the USA
Monee, IL
04 November 2026

67387016R00030